C000219244

1 MONTH OF
FREE
READING

at

www.ForgottenBooks.com

By purchasing this book you are eligible for one month membership to ForgottenBooks.com, giving you unlimited access to our entire collection of over 1,000,000 titles via our web site and mobile apps.

To claim your free month visit:
www.forgottenbooks.com/free296071

* Offer is valid for 45 days from date of purchase. Terms and conditions apply.

ISBN 978-0-265-68646-1
PIBN 10296071

This book is a reproduction of an important historical work. Forgotten Books uses
state-of-the-art technology to digitally reconstruct the work, preserving the original format
whilst repairing imperfections present in the aged copy. In rare cases, an imperfection in
the original, such as a blemish or missing page, may be replicated in our edition. We do,
however, repair the vast majority of imperfections successfully; any imperfections that
remain are intentionally left to preserve the state of such historical works.

Forgotten Books is a registered trademark of FB &c Ltd.
Copyright © 2018 FB &c Ltd.
FB &c Ltd, Dalton House, 60 Windsor Avenue, London, SW19 2RR.
Company number 08720141. Registered in England and Wales.

For support please visit www.forgottenbooks.com

UNITED STATES DEPARTMENT OF AGRICULTURE

BULLETIN No. 783

Contribution from the Bureau of Entomology
L. O. HOWARD, Chief

Washington, D. C.   PROFESSIONAL PAPER   July 14, 1919

THE RICE MOTH.

By F. H. CHITTENDEN,

Entomologist In Charge of Truck-Crop Insect Investigations.

CONTENTS.

INTRODUCTION.

Among the insect enemies of stored products which have been observed recently in this country, a small whitish larva or caterpillar of the moth *Corcyra cephalonica* Stainton (Pl. I) has attracted attention by its injuries. It resembles somewhat the better-known fig moth (*Ephestia cautella* Walk.). It has not been noted as a pest of importance, and has been given no common or English name. As it is somewhat widely reported as destructive to stored rice it may be called the rice moth. Beginning with October, 1911, complaints of damage by this insect were received from a firm manufacturing chocolate in western Pennsylvania, and a year later from another manufacturing firm in the same State, but the species was not positively identified until 1916.

NATURE OF INJURY.

The first correspondent of the Bureau of Entomology who wrote of this insect stated that beans of cacao (*Theobroma cacao*) imported from the Tropics were subject to attack by the larva. Apparently it laid its eggs in the beans, which are sometimes warehoused for several months, in the country from which they were shipped. During this period of storage additional generations of larvæ are hatched which destroy large quantities of the cacao beans or render them unfit for sale. The rice moths have been found most numerous in the older beans and also occur abundantly in cocoa nibs, in cocoa in powdered form, in refuse cocoa dust, and in ground cacao shells, so that they may be said to feed on any form of the cacao bean from the shells to the finished or edible article, cocoa or chocolate in powder, in cakes, and in confections, whether the substance is sweetened or unsweetened.

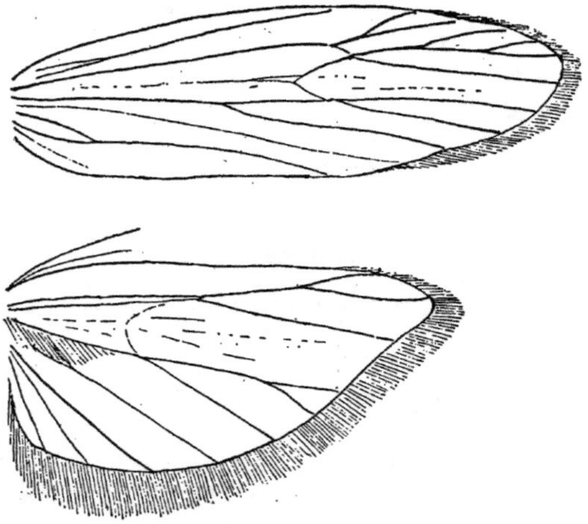

FIG. 1.—Diagram showing wing venation of the rice moth (*Corcyra cephalonica*). (After Durrant and Beveridge.)

Later moths and larvæ of this species were received in rice from different sources which will be mentioned hereafter.

This species works in much the same manner as do the fig moth (*Ephestia cautella* Walk.) and the Indian-meal moth (*Plodia interpunctella* Hbn.), forming a still stronger thread than do these related forms, and matting the infested material more closely. Indeed, this thread or webbing in the case of powdered cocoa becomes so dense that in close quarters the moths when emerging are scarcely able to make their exit. As a consequence of this and of the further fact that the food supply becomes too dry to be eaten, many of the larvæ perish. This is true not only under artificial conditions in the laboratory but has been noted in manufacturers' storerooms.

DESCRIPTIVE.

THE MOTH.

While, as previously stated, the rice moth resembles in certain respects some of our common moths which breed in stored cereals, dried fruits, and similar material, it does not belong to the same lepidopterous group, being a member of a different family, the Pyralidae, and subfamily, the Galleriinae, and closely related to a small group of moths which are best known as occurring in the combs of and Beveridge:

Antennae whitish fuscous; basal joint with some darker fuscous scales. *Head* and *Thorax* very pale fuscous, sometimes whitish fuscous, or darker fuscous. Fore wings very pale fuscous, the veins more or less indicated by darker fuscous scaling, and with a tendency to suffusion over the whole wing, except along the dorsum which remains of the pale ground-color; in some specimens the darker markings are almost absent, in others there is a tendency to form two irregular transverse dark lines, one at the end of the cell, the other at about half the wing-length, with some dark shading toward the base; a more or less distinct dark spot occurs on the margin at the end of each vein; cilia pale fuscous, with some admixture of darker scales. *Exp. al.* 14–24 mm. Hind wings, ♂ fuscous; ♀ shining whitish fuscous; cilia with a slightly paler line at their base. *Abdomen* and *Legs* pale fuscous.

SYNONYMY.

Corcyra cephalonica Staint., Ragonot, Ent. Mo. Mag., v. 22, p. 22, 23, 1885.
 Melissoblaptes (?) *cephalonica* Staint., Ent. Mo. Mag., v. 2, p. 172–173, 1866.
 Melissoblaptes translineella Rag.-Hamps., Mém. Lep., p. 491, pl. 45, fig. 23; pl. 51, fig. 26, 1901.
 Tineopsis theobromae Dyar, Ins. Inscit. Mens., v. 1, no. 5, p. 59, 1913.

THE EGG.

Pl. II.

The eggs have a pearly luster, are variable in shape, and have at one end usually a decided nipple, somewhat like that of certain fruits. The eggs are sufficiently large to be readily seen without the aid of a lens, and resemble somewhat those of the Mediterranean flour moth (*Ephestia kuehniella* Zell.). The exact dimensions have not been obtained.

THE LARVA.

Pl. III, A.

The larva when fully developed bears some resemblance to that of *Plodia interpunctella*. The sutures of the joints are somewhat more pronounced; the general color varies from white to a dirty, slightly bluish gray with occasional faint greenish tints. This dirty appearance of the larvæ is due to the dark material on which they feed and is especially evident in the immature stages. Larvæ which have fed on rice are more nearly white than those which develop from cacao preparations.

The head, without the mandibles, is truncate anteriorly and subtruncate posteriorly. The general color is rather dark honey-yellow, inclined to brown. The thoracic plate is pale honey-yellow, well divided at the suture and, while a little darker on the outer margin, is nearly uniform in color. The anal plate is very pale, scarcely darker than the joints. The three pairs of fore legs are rather long and prominent. The prolegs, with the anal legs, are also prominent but shorter. Observed under a strong lens the spiracles and piliferous tubercles are minute but distinct, and the pubescence, although sparse and of fine texture, is rather long, some hairs being nearly as long as the width of the body.

The average length when extended is about 13 mm. and the greatest width about 1.5 mm.

THE PUPA.

Pl. 111, B, C.

In general appearance the pupa resembles that of other cereal-feeding moths. The general color is pale yellow. The form is robust, and the arrangement of the segments is well shown in Plate III, B and C, the latter illustrating the ventral arrangement of the legs and wing pads. These latter extend nearly to the antepenultimate abdominal segment. The eyes, in fresh specimens, show merely as circular areas but when nearing transformation they become black. The antennal sheathes slightly overlap on the posterior margin. The best characters appear on the dorsum, the short median parallel elevated longitudinal lines evidently being characteristic, as they are nearly black and quite distinctly marked. The spiracles are small but distinct. The anal segment bears at the apex four processes, the anterior ones being in the nature of short spines.

Naturally there is a difference in the proportions of the pupa of this species as in the adult, the length varying from 7.5 mm. to 9 mm.

When about to transform the larva prepares a cocoon by joining together, by means of silken threads, a mass of the material on which it is feeding, as shown in Plate IV, A. An exposed cocoon is illustrated in Plate IV, B.

DISTRIBUTION.

While *Corcyra cephalonica* is known to occur in portions of Europe, Asia, Africa, and southern and insular America, it is by no means truly cosmopolitan. Durrant and Beveridge (9)[1] record the Mediterranean region, India and Ceylon, the Cocos Keeling Islands, Christmas Island, the Kei Islands, western Sudan, Nyassaland, La Réunion, Pará, Brazil, and Cuba and Grenada, West Indies. Ragonot (7) records Italy, the Ionian Islands, and the Seychelles. To this list may be added Porto Rico, Mexico, Hawaii, and Pennsylvania.

FOOD HABITS.

According to the authors just mentioned the rice moth would appear to be of eastern origin, introduced into Europe and elsewhere by the rice trade, and this is undoubtedly true. They further state

---

[1] Figures in parentheses refer to " Literature cited," p. 14.

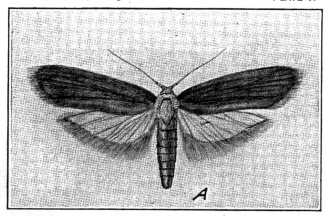

THE RICE MOTH (CORCYRA CEPHALONICA).

A, Mature moth; *B*, same in natural position at rest.   Much enlarged.

PLATE II.

EGGS OF THE RICE MOTH. HIGHLY ENLARGED.

PLATE III.

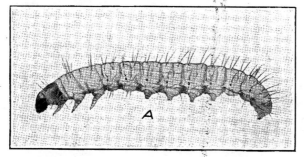

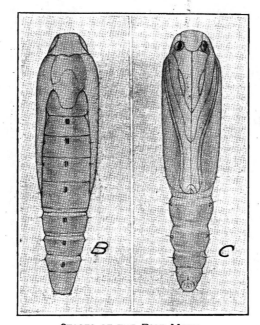

STAGES OF THE RICE MOTH.

A, Larva; *B*, pupa, dorsal view; *C*, same, ventral view. Much enlarged.

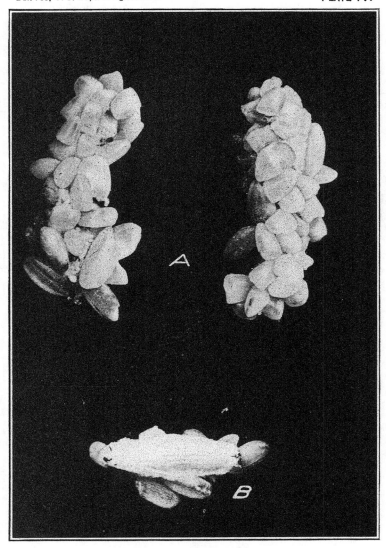

COCOONS OF THE RICE MOTH.

A, Exterior, showing grains of rice; *B*, cocoon exposed by removal of rice grains. Enlarged.

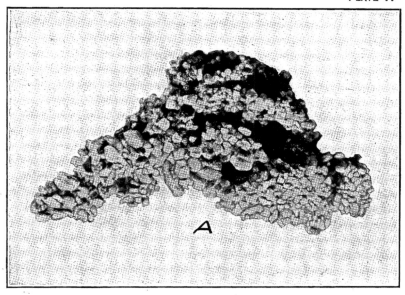

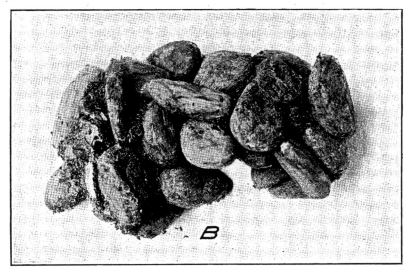

WORK OF THE RICE MOTH.

A, Mass of rice closely matted together by larvæ; *B*, cacao beans similarly attacked.

that it was thought to be especially attached to currants, that it is imported into England with Rangoon rice, which seems to be its natural food, and that there is little doubt that anything that will suffice for the genus Ephestia will be equally nourishing to the present species. This insect was also obtained in tins of army biscuit, but no particulars are given as to its breeding habits beyond what has already been said. The larva has been observed in Paris in the grain of sesame (*Sesamum orientale*) from Sudan, West Africa.

Plate IV and Plate V, A, illustrate the manner in which the cocoons of the rice moth are made by the larva in confining the

Fig. 2.—Army biscuit showing holes eaten by larvæ of the rice moth and webbing by same. (After Durrant and Beveridge.)

grains of rice by means of silken threads. Text figure 2 shows injury by the larvæ to an army biscuit, and Plate V, B, injury to cacao beans.

REPORTED INJURIES.

One of the firms which experienced trouble from this pest stated that the raw cacao beans, when received in bags, are stored in rooms about 16 feet high, some of the bags being piled nearly to the ceiling and others about 8 feet high. When the bags are disturbed the moths fly from between them and on examination numerous larvæ and cocoons may be found in such locations. Cocoons occupied or empty may be observed in almost any crevice in the walls of the storerooms.

Correspondents also note that the oldest cacao beans are, as a rule, the most heavily infested.

May 6, 1914, 10 moths of this species were placed in a rearing jar with cacao beans as food. One moth was still alive on May 27, but was found dead the following day, having lived 21 days without food. According to Dyar the tongue is completely absent in this moth, so it is unable to feed. No evidence of insect attack could be noted through the glass jar when examined on July 9, but when some of the beans, which had become moldy on account of the moist weather during this period, were opened, a mature larva and a cocoon containing a pupa were found. Attack was confined chiefly to beans that already had been injured more or less.

March 8, 1916, Dr. Carl Michel, United States Public Health Service, San Juan, P. R., furnished moths and pupæ, the latter in webbed-up rice, and stated that the species infests warehouses in Porto Rico, that the eggs are laid in sacks of cereals, and that the developing larvæ render the cereals unfit for human consumption. The merchants at San Juan claim that the rice is infested before it reaches that port and that nearly all of it is concentrated at New Orleans or Galveston for shipment. The claim is not made, however, although it is inferred, that the insect is shipped from the United States, but it seems more probable that the moth has been established in Porto Rico for a number of years. Agents of the Bureau of Entomology spent much time from 1908 to 1916 investigating insects injurious to rice and other stored products from New Orleans and Galveston, but they did not observe this insect at these or other ports. It may have been introduced recently through carelessness in vessels returning from Porto Rico containing foodstuffs on which it was able to subsist. On March 22 Dr. Michel sent additional specimens of larvæ in infested rice. The larvæ were all paler than were those reared from darker substances, such as chocolate and similar products, and as a result it was noted that the piliferous tubercles were plainly visible, whereas in the darker forms they were scarcely noticeable. September 12, 1916, numerous larvæ and some pupæ of this species were received in rice from the same source. The correspondent stated that some of the moths had been breeding continuously since the previous February, and that they thrived at room temperatures.

May 19, 1916, samples of rice infested by this species were again received, and on September 18 the Bureau of Chemistry reported that this specific shipment of rice was California grown, milled in San Francisco, and shipped via Panama Canal to New York City where it was held for about 30 days, and then reshipped to San Jaun, P. R., where upon its arrival the buyers rejected it because the market had declined, but not on account of "vermin," as the rice was apparently in sound condition. The rice was kept until October

30, and in the meantime the rice moth and other pests developed and the rice was condemned by the United States Government. Finally, the rice was shipped to New Orleans to be reconditioned, and was put into a condition satisfactory to the Federal authorities.

December 7 of that year, a chocolate firm in Pennsylvania, which previously had furnished specimens, wrote that the moths disappear with the arrival of cold weather and are not seen again until the following spring. During the late spring months and all summer they are in evidence. The greatest trouble is experienced from the laying of eggs by the moths on the finished chocolate and cocoa. The eggs hatch into larvæ and the customer naturally objects to " wormy " goods. Attempts were being made to avoid this as much as possible by keeping finished materials covered.

LIFE HISTORY.

The complete life history of the rice moth has not been ascertained. The progress which might have been made with other insects in similar investigations was prevented in this case by the fact that seldom more than two generations were obtained in a single rearing jar of cocoa or related substances. When confined in large numbers the larvæ, like others of similar habits, such as Ephestia, travel, evidently in an endeavor to secure a suitable location for transformation to pupæ, to a greater extent than do the other species. This might explain the fact that the pupal cases or cocoons usually are found either on the outside of the bags at point of contact in the piles, or in the folds of the burlap sacks, which provide more or less shelter. In the rearing jars, although small pieces of cloth were inserted to form shelters for the pupæ, the thick webbing spun by the larvæ completely covered the infested material, preventing the exit of the moths, which died without being able to reproduce. This fact is mentioned because it happened in the case of a half dozen rearing jars of large size (about 8 liters capacity).

It has been ascertained, nevertheless, that the insect requires only a short time to develop from larva to adult, this period being dependent on temperature. The entire summer period for transformation from egg to egg is between 28 and 42 days, or from 4 to 6 weeks, but this period would be prolonged considerably in cooler weather.

Better results attended rearing experiments with this species in infested rice from Porto Rico. From a lot of moths which deposited eggs about May 26 a new generation of moths began to issue July 8, this period having been passed in 43 days, or approximately 6 weeks. The temperature ranged from 52° to 82° F., reaching the maximum only on a few occasions, and the average or mean temperature for the experiment was from 68° to 70° F.

The question has been raised by importers and manufacturers as to whether or not it is possible to retard the development of the rice moth in order that control measures may be undertaken at desirable times. While it was not possible to undertake any experiments along this line, it is known from analogy that development could be considerably retarded by cold storage. The egg period might be extended from the normal length of time, 3 to 10 days, to about a month; the larval period to 6 months or more; and the pupal period from the normal of from 5 to 14 days to 4 weeks or longer, making a possible total of about 8 months.

While complete life-history data of this species would be desirable, what has been learned is sufficient to show that such life-history studies would not differ essentially from those of related species, such as the Mediterranean flour moth and the Indian-meal moth, and it has been developed that there is a practical certainty of four generations annually and a possibility of as many as six in high temperatures.

ASSOCIATED INSECTS.

The fig moth (*Ephestia cautella* Walk.), as previously stated, has been associated with this species in infested rice and cocoa products. In one rearing jar containing the rice moth breeding in cocoa, received June 18, 1915, the larvæ of the latter were full grown on August 27. The jar was examined again on September 10 and apparently contained only the fig moth with its larvæ. This latter had evidently " run out " the former, its larvæ perhaps feeding on the larvæ and pupæ of the rice moth, which in nature is not an unusual occurrence.[1] Some, however, remained, and in a few days the rice moth reappeared. In this particular rearing cage the fig moth must have deposited her eggs through the mesh covering the jar, although this was decidedly thick and closely woven. Fig-moth females have been known to do this in previous instances.

The Indian-meal moth (*Plodia interpunctella* Hbn.) developed in great numbers in a lot of chocolate in which the rice moth had been reproducing abundantly, completely devouring the edible material and then perishing.

It may be noted that when closely confined with edible material the three moths mentioned, in common with others which feed upon stored products, frequently perish because of the compact webbing which prevents escape and the lack of moisture which produces excessive drying of their food supply, curtailing the longer reproduction period of the species.

---

[1] The larvæ of the cabbage worm (*Pontia rapae* L.) have been noted feeding on the eggs of the cabbage looper (*Autographa brassicae* Riley). The corn earworm (*Chloridea obsoleta* Fab.) is also well known to be cannibalistic.

Some forms of beetles, however, are able to continue feeding in the absence of moisture until the supply of food is exhausted.

The saw-toothed grain beetle (*Silvanus surinamensis* L.) has been found in several instances associated with the rice moth. Obviously it plays the same rôle with this species as with other moths—a scavenger, although a decidedly noxious pest.

The rust-red flour beetle (*Tribolium ferrugineum* Fab.) has been observed in the same situations as the saw-toothed grain beetle.

The lesser grain-borer (*Rhizopertha dominica* Fab.) was received in rice from Porto Rico associated with stages of the rice moth.

The Siamese grain beetle (*Lophocateres pusilla* Oliv.) was observed breeding in numbers in a sample of Porto Rican rice some time after receipt, showing that the immature stages were present at an earlier date.

The rice weevil (*Calandra oryza* L.) was present in small numbers in most of the samples inspected. It was noticeable in broken rice that the beetles which developed in such small quarters were not as large as those which are found in soft kernels of corn and wheat. The color of the beetles taken in broken rice was brighter and they had the appearance of being a distinct species.

HISTORY AND LITERATURE.

While the rice moth probably has been present in Europe for many years, it was not until 1866 that it was discovered in York, England, and described as a new species by Stainton (1). It was found in imported dried " currants " (*Passulae corinthicae*), called " Corinthian currants," but in reality a well-known species of grape. In 1875 Barrett (2) mentioned the occurrence of this species in fruit warehouses in London, together with other insects of similar habits. In 1885 (3), 1893 (4), and 1901 (7) Ragonot wrote, in technical articles, in regard to the classification and characters of this species, without reference to its injurious habits. In 1895 Meyrick (5) gave a brief technical description of the adult, stating that the larva occurs in dried " currants." In 1897 (6) the author mentioned this species in a list of insects likely to occur in this country in dried fruit. In 1909–10 Fletcher (8) recorded the species as occurring in rice from the West Indies.

In 1913 Durrant and Beveridge (9) wrote the most extensive account of the insect which had appeared to that date, referring especially to its occurrence in army biscuits and the temperature which would destroy this and other species of related habits. An article dealing with this insect, by Otto H. Swezey (10), appeared the same year.

In 1908 the rice moth came to the attention of Mr. Jacob Kotinsky of the Bureau of Entomology, at that time in Hawaii, who found it breeding in a feed warehouse in Honolulu in July. On July 10, 1909, it was captured at Kaena Point by Mr. Swezey. The latter part of the same month moths were found emerging from a package of cracked wheat obtained from a Honolulu grocery. Mr. Swezey expressed the opinion that although the species is a European moth apparently not recorded at that time in the United States, it certainly must have reached Honolulu from the United States.

The habits of the moth are well described by Barrett (2). He states that when disturbed in flight, unlike Ephestia and Plodia, it darts down in a zigzag and almost immediately comes to rest. Toward evening the males run about, quivering their wings in a peculiar manner. The moth shows considerable skill in selecting for a resting place the projections of rough beams, to which, owing to its rough, blunt head and closely folded wings, it bears so close a resemblance that Barrett states he has taken specimens between his fingers before he could satisfy himself that they were not projecting splinters. This can be readily appreciated by reference to Plate I, B, which shows the moth at rest. The moth is peculiarly sluggish, even more so than those of the other genera. Barrett writes of this and of a related species (Ephestia) that they were being replenished constantly from imported dried fruits, since every cargo of fruit swarmed with the larvæ, some of which died from change of climate and other causes, but many of which came to maturity. He states that it is obvious that places in which old "currants" have been stored are the most potent sources of infestation, the new fruit coming into harbor during the month of September when the moths are already plentiful. He believed that the different species occurred in about equal numbers and was certain that they had formed a settlement from which it would be no easy task to expel them.

CONTROL MEASURES.

Warehouses and other structures in which the rice moth has become established should be cleansed thoroughly. Any bags which contain or have contained infested rice or other cereal, cacao beans, cocoa or similar material, or dried fruits should be fumigated; all corners, cracks, and crevices which may harbor the insect should be brushed out; and all refuse promptly destroyed by burning. The walls and floors then may be washed down with a soluble creosote disinfectant, or a solution of common salt. The brushes used should be stiff and strong, and every point should be reached so as to make the compartment perfectly clean.

The machinery also should be cleaned thoroughly and the entire plant fumigated with hydrocyanic-acid gas. In small plants either carbon disulphid or sulphur dioxid may be employed for fumigation, but if the buildings are so constructed that heat of 120° to 130° F. may be applied for several hours, the same result will be accomplished.

Secondhand bags should not be used without first disinfecting them and bags previously used for the transportation of cacao beans or other food materials which the rice moth is known to attack should be examined for the presence of the insect in its various stages. When insects are found it is best to establish a quarantine bin, room, or fumigator in which the infested bags may be thoroughly baked or fumigated before they are taken into the main building. If it is desired to fumigate a compartment containing bags filled with cacao beans, rice, or similar material the bags should first be brushed off carefully and the tiers of bags so separated as to leave air space between in order that the gas may penetrate the contents more readily. Even after fumigation there is always a possibility that a small percentage of the insects may remain and revive.

DESTRUCTION BY HEAT.

Treatment of insect-infested stored products by heat is by no means a new remedy, but large-scale work with this method had not been conducted to any extent until about 10 years prior to the time of writing. This method appears to have been first successfully used in the control of mill pests at that time by the Kansas Agricultural Experiment Station, for the control of both the Mediterranean flour moth and the Indian-meal moth. Soon thereafter Mr. C. H. Popenoe, of the Bureau of Entomology, conducted experiments in Virginia, under the writer's direction, which were quite successful against both of these pests.

The heat method is equally applicable for the rice moth, although it is valuable only for mills or other structures heated or operated by steam, since it presupposes the installation of necessary heating pipes and radiators. The temperature required, from 120° to 130° F., can be obtained readily in a mill provided with sufficient radiation surface to maintain a winter temperature of 75°. A warm, quiet day should be selected for best results, and the temperature after being reached should be maintained for 8 hours or more in order to insure penetration. Should additional radiation surface be required, it may be provided by the installation of temporary supplementary coils of 1¼-inch pipe, which will operate to best advantage if placed near the floor. In mills where a complete installation is required, radiators should be calculated on a basis of 1 foot of heating surface (2⅓ linear feet of 1¼-inch pipe) to from 50 to 100 cubic feet of space, depending on the construction of the building and the situation of the coils. The maximum figure should be applied to the lower floors.

A steam pressure of from 75 to 100 pounds may be employed advan-tageously. Since bags of compact material are heated to the center with difficulty, so far as possible they should be separated before treatment to facilitate uniform heating, for insects and their larvæ become more active upon the application of the heat and may work their way to the center of the bags in their efforts to escape it.

Better results may be obtained by providing the radiators with water traps or vents.

Rice and cacao beans should not be exposed to a temperature above 130° F. for more than one hour, as excessive splitting takes place in rice, especially if bleached, and, owing to the excessively oily nature of cacao beans, they may become rancid.

Germination in the case of some seeds, such as peanuts, is not af-fected even by an exposure of six hours to a temperature as high as 140° F., but it is best to be on the safe side in the treatment of com-modities affected by this moth until we have had more experience along this line. It should be added that a temperature of 140° F. is fatal to most forms of insect life in a short time—larvæ, pupæ, and adults. The Indian-meal moth, it has been learned by experiment in the Bureau of Entomology, dies in less than half an hour when so exposed.

FUMIGATION METHODS.

HYDROCYANIC-ACID GAS.

For the fumigation of buildings and other structures inhabited by the rice moth, the hydrocyanic-acid gas process is the most useful. Indeed, it is now the standard remedy for practically all insects affecting stored products. It has been in use for this purpose for about 20 years and most progressive millers are familiar with the method of application. Information in regard to hydrocyanic-acid gas fumigation has been furnished by the Bureau of Entomology in various bulletins and other publications. In the earlier ones the use of cyanid of potash or potassium cyanid was advised, but owing to conditions brought about by the war it is now impossible to secure this chemical, and as a result cyanid of soda or sodium cyanid is being used, and while somewhat expensive, is much cheaper than the corresponding potash salt. The formula is as follows:

```
Sodium cyanid_____avoirdupois ounce__  1
Sulphuric acid_____fluid ounces__  1½
Water _____do____  3
```

Information in regard to this method is furnished in Farmers' Bulletin 699, "Hydrocyanic-acid Gas Against Household Insects." While this, as the title shows, is especially for dwellings, the methods advised can be adapted readily to mills and storehouses.

Hydrocyanic-acid gas, it must be stated, is the most poisonous substance in common use, but it is still employed very extensively in fumigating mills and dwellings, and if the directions in the bulletins cited are carefully carried out there is really no danger to human beings.

CARBON DISULPHID.

Before the general adoption of hydrocyanic-acid gas as a means of fumigating buildings, carbon disulphid was considered a standard, and it is still of value, particularly on a small scale, as a substitute for hydrocyanic-acid gas. It is extremely inflammable, however, which has led to its abandonment in many localities. Directions for its use are given in Farmers' Bulletin 799 [1] " Carbon Disulphid as an Insecticide."

SUMMARY.

1. The rice moth (*Corcyra cephalonica* Staint.) has been known to occur in the United States only since 1911, and was not identified until 1916.

2. Its origin is unknown, but it has been introduced at many points in other continents and is as yet not strictly cosmopolitan. It has been found commonly in England, where it was introduced in rice, chiefly from India and Burma, and also in dried fruits.

3. Its habit of feeding on cacao beans is probably an acquired one. Evidently it is inclined to be omnivorous, since it breeds in rice, dried fruits, the various products of cacao, such as cocoa, cacao shells, and sweetened and unsweetened chocolate, ship biscuits, and sesame seeds. It displays, however, no partiality for any of these food substances.

4. Its complete life history has not been traced, but, like other indoor species, it reproduces nearly the year around under average conditions. In the United States infestations appear to die down from time to time, but are stimulated through new shipments of cacao beans from South America and Central America.

5. It produces copious and dense external webbing to which food materials, such as rice, cocoa, and other matter, strongly adhere. In this respect its work and injury resemble those of the fig moth (*Ephestia cautella* Walk.) and related species, and the Indian-meal moth (*Plodia interpunctella* Hbn.).

(6) While it has been recognized only from western Pennsylvania and Porto Rico, it occurs without doubt at other points, and dealers in rice, chocolate, and similar imported dry edibles should keep a

---

[1] The Farmers' Bulletins mentioned may be obtained free on application to the Division of Publications, United States Department of Agriculture.

lookout for the species to prevent it from gaining entrance and be-coming established in large warehouses and similar plants.

7. It will undoubtedly increase in injuriousness in time unless proper measures are taken to stamp it out by thorough treatment.

LITERATURE CITED.

(1) STAINTON, H. T.
   1866. Description of a new species of the family Galleridae. *In* Ent. Mo. Mag., v. 2, 1866, p. 172–173.

   Original description as *Melissoblaptes* (?) *cephalonica* n. sp., from York, Eng., from dried " currants " (imported).

(2) BARRETT, C. G.
   1875. On the species of Ephestia occurring in Great Britain. *In* Ent. Mo. Mag., v. 11, p. 269.

   Page 272 : In fruit warehouses in dried " currants." Notes on habits.

(3) RAGONOT, E. L.
   1885. Revision of the British species of Phycitidæ and Galleridæ. *In* Ent. Mo. Mag., v. 22, 1885–6, p. 17–32.

   Pages 22–23 : Remarks ; placed in genus Corcyra from the country of its supposed origin.

(4) ———.
   1893. Monographie des Galleriinae et Phycitinae. *In* Romanoff, N. M., Mémoires sur les Lépidoptères. t. 7. Saint Petersbourg.

   Illustrations : Head of female, pl. 1, fig. 34 ; venation, pl. 3, fig. 18.

(5) MEYRICK, E.
   1895. A handbook of British Lepidoptera. 843 p., illus. London.

   Page 384 : Technical description of the moth and brief notes.

(6) CHITTENDEN, F. H.
   1897. Some little-known insects affecting stored vegetable products. U. S. Dept. Agr. Div. Ent. Bul. 8, n. s. 45 p., 10 figs.

   Page 10 : Mere mention as a species likely to be found in this country in dried fruit.

(7) RAGONOT, E. L.
   1901. Monographie des Galleriinae. *In* Romanoff, N. M., Mémoires sur les Lépidoptères. t. 8, p. 421–507. Saint Petersbourg.

   Pages 491–493, pl. 45, fig. 23, and pl. 51, fig. 26 : Definition of genus Corcyra, description of *cephalonica* and *translineella* (=synonym) and plate of each.

(8) FLETCHER, T. B.
   1909–10. Lepidoptera, exclusive of the Tortricidæ and Tineidæ, with some remarks on their distribution and means of dispersal amongst the islands of the Indian Ocean. *In* Trans. Linn. Soc., s. 2, v. 13, Zoology, p. 265–323, pl. 17.

   Pages 296 and 316 : Recorded from West Indies ; mention as common in rice stores.

(9) DURRANT, J. H., and BEVERIDGE, W. W. O.
   1913. A preliminary report of the temperature reached in army biscuits during baking, especially with reference to the destruction of the imported flour-moth, Ephestia kühniella Zeller. *In* Jour. Roy. Army Med. Corps, v. 20, no. 6, p. 615–634, 7 pl.

   Pages 633–634 : Occurrence in army biscuit, description, bibliography, and distribution ; illustrations of the moth, larva, and injury.

(10) SWEZEY, OTTO H.
    1913. Notes on two galleriids. *In* Proc. Hawaiian Entom. Soc., Honolulu, Hawaii, v. 2, p. 211–212.

        Page 212: Occurrence in Hawaii in 1908–1909 in a feed house and in cracked wheat.

(11) DYAR, H. G.
    1913. A galleriine feeding in cacao pods. *In* Ins. Inscit. Mens., v. 1, no. 5, p. 59.

        Characterization of *Tineopsis* n. g., and description of *T. theobromae* n. sp., as follows: "Dark gray; fore wing without markings. Hind wing paler, silky gray. The head is heavily tufted and with the narrow, pointed wings gives the insect the aspect of a Tineid. Expanse, 13–15 mm."

ADDITIONAL COPIES
OF THIS PUBLICATION MAY BE PROCURED FROM
THE SUPERINTENDENT OF DOCUMENTS
GOVERNMENT PRINTING OFFICE
WASHINGTON, D. C.
AT
10 CENTS PER COPY
▽

WASHINGTON : GOVERNMENT PRINTING OFFICE : 1919

CPSIA information can be obtained
at www.ICGtesting.com
Printed in the USA
BVHW040802081218
535028BV00023B/63/P